Pedro S. Pontes de Oliveira

Automatic irrigation for pot plants

Pedro S. Pontes de Oliveira

Automatic irrigation for pot plants

ScienciaScripts

Imprint

Any brand names and product names mentioned in this book are subject to trademark, brand or patent protection and are trademarks or registered trademarks of their respective holders. The use of brand names, product names, common names, trade names, product descriptions etc. even without a particular marking in this work is in no way to be construed to mean that such names may be regarded as unrestricted in respect of trademark and brand protection legislation and could thus be used by anyone.

Cover image: www.ingimage.com

This book is a translation from the original published under ISBN 978-613-9-66563-1.

Publisher:
Sciencia Scripts
is a trademark of
Dodo Books Indian Ocean Ltd. and OmniScriptum S.R.L publishing group

120 High Road, East Finchley, London, N2 9ED, United Kingdom
Str. Armeneasca 28/1, office 1, Chisinau MD-2012, Republic of Moldova, Europe
Printed at: see last page
ISBN: 978-620-8-03642-3

ACKNOWLEDGEMENTS

I dedicate this work to my friends who had a lot of patience with me throughout its structuring. They helped me with my doubts, encouraged me and at times gave up their own time to help me. Of these friends, Alisson and especially Luís Ilderlandio (Lucas) were always willing to help.

Another person who was very important to the completion not only of this work, but of the course as a whole, was Paulo Peixoto, a great friend who was always there to encourage me and prevent me from becoming discouraged in the face of difficulties.

To Prof Vanier de Andrade, who helped with excellent practical guidance and was always willing to share his knowledge.

To Prof Saulo Lima Bezerra, who believed, encouraged and guided us, offering great suggestions that influenced the improvement of the work as a whole.

To my siblings, Edilce Stefanie Pontes de Oliveira and Gleydson Henrique Pontes de Oliveira. Especially my parents, Jose Cavalcante Pontes and Maria Hilda de Oliveira Pontes, who were essential throughout my journey. All of them were always united, supporting us and thus making it easier to overcome the obstacles that arose, and this union was the greatest inspiration for me to never give up on my goals.

Kaizen (from the Japanese, change for the better) is a word of Japanese origin with the meaning of gradual continuous improvement in life in general (personal, family, social and work).

"Today better than yesterday, tomorrow better than today!"

Author unknown.

SUMMARY

Electronics play a fundamental role in modernising various activities. It makes it possible to automate equipment, making activities faster and more precise. Most technological innovations are based on electronic devices. Over the last few years, electronics has undergone an extremely rapid evolution, occupying a prominent position among all other technologies (BRAGA, 1999; QUEIROZ, 2007).

Irrigation is no different. Personal, financial and natural resource needs are becoming better utilised every day due to the adaptations and improvements coming from automation. This work explains the real needs of plants grown in restricted containers (pots), showing important information and the correct way to irrigate them. Using electronic equipment developed with targeted adaptations, creating the most favourable growing environment possible for the development and perfect health of the plant. Making it viable and offering greater comfort for people who don't have the time or knowledge for this type of cultivation.

Key words: automation, irrigated agriculture, electronics, soil.

SUMMARY

CHAPTER 1	5
CHAPTER 2	7
CHAPTER 3	8
CHAPTER 4	21
CHAPTER 5	30
CHAPTER 6	31
CHAPTER 7	32

CHAPTER 1

INTRODUCTION

Growing plants in small containers is the solution for those who don't have a plot of land but like to keep greenery close at hand. The secret to having beautiful and healthy plants at home is to give them conditions close to those of their original habitat. In other words, think about the composition of the soil, the incidence of light, water and nutrition.

To find out how much water a plant needs, you need to observe the development of the plants to discover their needs. Most of the time, plants die from too much water, not too little. The ideal is to water frequently and not too much. Waterlogged soil favours the appearance of fungi, pests and causes the roots to rot. The absence of water, on the other hand, causes the roots to dehydrate and the plant to die.

Plants that live in the wild develop their roots at great depths until they find the water stored in the soil. This is not the case with plants grown in pots, which is why they need to be watered more or less frequently depending on factors such as the type of soil, the size and depth of the container, where it is exposed to the sun's rays, the species of plant and so on. There is no mathematical rule to determine how often you should water a potted plant. Watering should be done as often as necessary. A simple way to realise this is to touch the soil with your fingers and feel if it is dry. If so, you should water it immediately. If it's damp, there's no need to water. We mustn't confuse **"damp soil" with "soggy soil"**, the latter suffocates the roots, causing them to rot and the plant to die. This is one of the main causes of failure for many growers, so we must avoid these two situations as much as possible (IUBlOG, 2011).

Seeing the need for time and dedication to ensure the health of plants grown in pots of different sizes. Especially the shallow and larger diameter ones, which arrange the soil in such a way as to favour moisture loss through evaporation. And because of a little distraction or forgetfulness, the soil can remain dry for a long time. Methods such as drip irrigation prevent the soil from drying out, but can generate a very high level of humidity, which is also inappropriate and can cause the death of the plant in both situations. Analysing this fact, we came up with the idea of creating a device that meets specific needs, being able to measure the ideal humidity in the soil, the level of water in the container, and using this information being able to manage it automatically using a microcontrolled system, sensors and actuators (pump motor), informing the variations in the whole process visually via LEDs. This will increase the time available for people who don't have the time to practise this type of cultivation. By using the equipment, we can avoid unsuitable conditions, optimise the rational use of water and, in the case of bonsai growers, avoid the loss of decades of dedication to growing these plants, since there are bonsai that are centuries old and of inestimable value.

This monograph will initially provide information on the importance of automation in processes, focussing on irrigation. Citing information on soil composition and its influences on water in its ideal proportion. Adding to the reader's knowledge and some characteristic precautions for this crop. Facilitating understanding of the principles used to structure this project as a whole.

1.1. Development Methodology

To realise the project in question, several stages of development were necessary:

- Survey of bibliographical references, through research into books, articles, publications, the internet, among others in areas related to the project;
- The selection of materials, both bibliographical and equipment used to carry out the work;
- A survey of the relationship between automation and irrigation;
- Study of the principles for automating a system;
- Definition of the soil's physical-chemical influence on its electrical conductivity;
- Methodology for preparing soil in pots;
- Carrying out tests and correcting errors;
- Analysis and presentation of results.

CHAPTER 2

OBJECTIVES

2.1. General Objective

- To develop a portable, low-cost irrigation device that integrates all the necessary components for automated irrigation, which illuminates all the stages of the process, giving the people who own it time, comfort and safety.

2.2. Specific objectives

- Research into automated irrigation systems;
- Studying the physical and chemical characteristics of the soil;
- Research into cultivation methods characteristic of plants grown in pots;
- Analysing the characteristics of humidity and magnetic sensors;
- Develop the microcontroller software;
- Carry out simulations to calibrate the sensor.

CHAPTER 3

LITERATURE REVIEW

3.1. Automation x Irrigation

Automation can be defined as a systematic method that, through the use of mechanical or electro-electronic devices, is capable of controlling machines and processes without human intervention (QUEIROZ, 2007). These devices can include computers or other logical devices (CNCs, Microcontrollers, etc.). It is applied to processes with the aim of making them more efficient, i.e. maximising production with lower energy consumption, better safety conditions, both human and material and the information inherent in the process.

According to Silveira (1998), in automation a distinction is made between rigid systems and flexible systems. The former are made up of electromechanical components such as relays and contactors. Signals from sensors coupled to the machine or equipment to be automated trigger relay logic circuits that drive loads and actuators. With the advance of electronics, they have become faster, more compact and capable of receiving more input information and acting on a greater number of output devices. Memory units have increased in capacity, allowing more information to be stored.

With the growing evolution of automation, irrigated agriculture, like other activities, needs to keep up with technological advances. In recent times, electronic devices and computers have made increasing technological progress, becoming cheaper, more precise, more manageable and with a better user interface. Irrigation automation can be achieved using a combination of **soil moisture sensors, PLCs,** microcontrollers, solenoid valves and a pressurised source of water via pumps or not. The aim is to apply the right amount of water at the right time, avoiding both lack of humidity and water stress, giving better results.

The advantages of using automatic equipment in a process over a human operator are generally: cost, attendance, speed and precision. It also has the disadvantage of being unable to react to disturbances and accidents not foreseen in its design.

3.2. Principles of Process Automation

Automation offers the possibility of making a process faster and more effective, as long as attention is paid to what is to be automated and how it is to be done. Identifying a problem or need is the first step to starting automation, i.e. describing what you want to automate. Next, analyse and understand the problems, thereby defining possibilities for solving them. Once this is done, it is necessary to define what will be given as input and output, foreseeing all possible situations and how this information will be measured so that it can be understood by the system (PELLISON, 2001).

3.2.1. Control System

A control system can be defined as a collection of physical components connected together. In such a way that it can control, direct or regulate itself or other systems. Through sensors, it can react by interpreting adverse stimuli or excitations. These external stimuli that are applied to a system are known as input, and the response obtained from a control system is called output (PELLISON, 2001).

Control systems are categorised into open-loop and closed-loop systems. In an open-loop system, the control action is independent of the output, while in a closed-loop system the control action depends in some way on the output, as shown in figure 01:

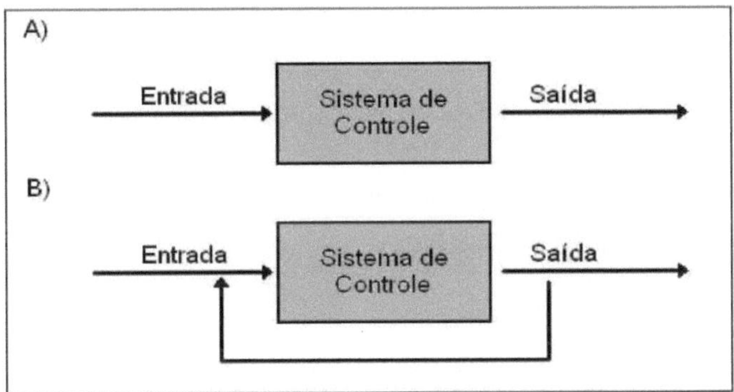

Figure 01- Control System: A) Open B) Closed

There are several factors that cannot be disregarded when it comes to defining a system's input data. In the case of irrigation, the main factor is the type and variations of the soil. This is because physical and chemical characteristics can influence electrical conductivity, which is one of the main factors used as input data by soil moisture sensors.

3.3. Soil

Soil is made up of mineral and gaseous elements such as clay, silt, stones and nutrients. Soil is also a living system, as it contains an immense number of life forms. The decomposition of plant remains in the soil, carried out by fungi, bacteria, earthworms and insects, results in the mineralisation of nutrients (carbon, nitrogen, phosphorus, sulphur, etc.), which are directly assimilated by plants or form other compounds.

Image01- Earthworms in the soil.

Source: http://www.sobiologia.com.br/conteudos/Morfofisiologia_vegetal/morfovegetal37.php

The solid fraction of the soil is in the form of a mixture of grains with varying shapes and sizes, which are classified according to their diameter into granulometric fractions. The coarse fractions correspond to the soil skeleton (particles with a diameter greater than 2 mm), which are gravel (from 2 mm to 2 cm in diameter), pebbles (from 2 cm to 20 cm) and shale (diameter greater than 20 cm). The smaller particles include the so-called "fine earth" (particles smaller than 2 mm in diameter), which includes sand (with a diameter of 0.05 mm to 2 mm), silt (from 0.002 mm to 0.05 mm) and clay (with a diameter of less than 0.002 mm) (MONIZ, 1975). Image 02 shows the granulometric variations of the soil and its layers respectively.

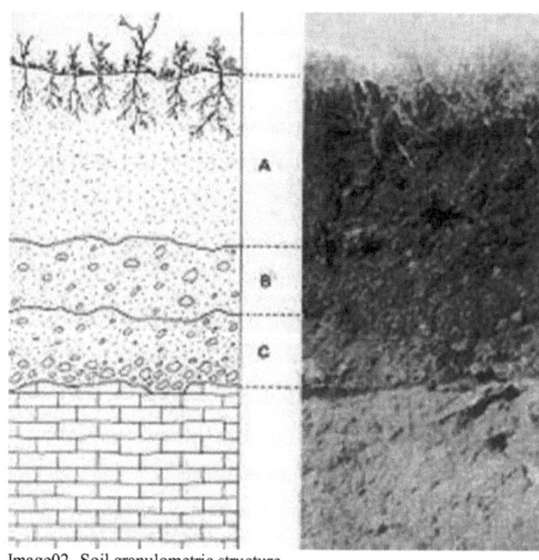

Image02- Soil granulometric structure.

The soil's pore space is occupied by air and soil water. Soil air provides the oxygen needed for the respiration of roots and microorganisms, and differs from atmospheric air in some chemical characteristics. Heavily compacted soils prevent the movement of air and water, causing various losses when it comes to growing plants (BRADY, 1989).

Water, together with inorganic and organic ions in solution, forms the soil solution. This is important not only as a source of water for plants, but also to enable roots to absorb the inorganic ions in the soil solution (BRADY, 1989).

Organic matter is a source of nutrients, with the ability to increase water retention, improve soil structure, porosity, reduce soil density, consistency, colour, among others (LOPES, 1989).

Soil composition can influence water retention capacity. In a sandy soil, there is faster infiltration and little water retention due to the pore space, which allows water to drain freely. Due to these characteristics, soils are naturally drier because they retain little water. They are loose and less prone to compaction than clay soils. In clay soils, on the other hand, there is greater water retention in the soil due to the presence of microspores that retain water against the forces of gravity, but these soils can be easily compacted. This further reduces the pore spaces, which limits the movement of air and water through the soil (LOPES, 1989).

3.3.1. Soil Water Retention

Water retention capacity is a characteristic that varies depending on the composition of the soil. In the words of LEPSCH (2002, p.12),

> Soil can retain water, storing it for a certain period of time. Plants use this water by absorbing it and, for the most part, returning it to the atmosphere in the form of vapour. In this way, the water absorbed in liquid form between the soil particles is emptied from the pore spaces. It can be replaced naturally by rainfall or artificially by irrigation. Inside the soil it can be retained both in the pores between aggregates and in thin films around the surface of colloidal particles. According to the content and nature of moisture retention, 3 soil states are recognised: a) wet; b) moist; c) dry.
>
> In wet soil, all the pores are filled with water and air, and air is practically absent. Under natural conditions, once all the pores have been filled with water and it ceases in the larger pores, it drains downwards, or laterally, wetting the deeper parts, or joining the underground water table and giving rise to springs. This water is called gravitational because it seeps into the ground under the action of gravity.
>
> Moist soil contains air in the macropores and water in the micropores. The smaller pores

act as capillary tubes and for this reason the water is referred to as capillary water. It is retained in the soil with such force that it can hold on even against the action of gravity, but this force is not so great as to prevent the roots from extracting it, and therefore represents storage available to plants. Not all soils have the same capacity to store water. It varies according to characteristics such as texture, structure and organic matter content. Sandy soils with little humus have a lower capacity to store water than clayey soils rich in humus.

Even when air-dried, soil can still contain a certain amount of water in the form of extremely thin films around the colloidal particles. This water is retained with a force greater than the extraction capacity of plant roots and is therefore called inactive water.

Soil water contains small and variable amounts of mineral salts, oxygen and carbon dioxide, thus forming a dilute solution known as the soil solution. The type and quantity of elements dissolved in this solution varies according to the type of soil, directly influencing electrical conductivity (LEPSCH, 2002).

3.3.2. Soil Electrical Conductivity

Electrical conductivity has attracted attention as a tool used in soil mapping, as it is a quick and economical method that indicates soil productivity (McBRIDE et al., 1990). According to Molin et al. (2005), in addition to yield maps, investigations related to soil electrical conductivity can be helpful, as they facilitate the measurement of clay content (WILLIAMS and HOEY, 1987), water content (McBRIDE et al.,1990), organic matter content (JAYNES, 1996), among others.

Electrical conductivity is the ability of a material to transmit (conduct) electrical current. It is a property of the material, just like density or porosity. Soil can conduct electric current through the water it contains, where it has dissolved electrolytes in its composition, and this solution is the main factor favouring this conductivity (MOLIN, 2005).

The method that uses electrical resistance for moisture sensors requires perfect contact between the soil and each electrode. This is necessary in order to introduce a current into the soil and measure the resulting voltage. By varying the distance between the electrodes, the resistance also varies proportionally, as this increases the difficulty of the current reaching the other rod or conductor.

This work uses a resistive sensor that is in direct contact with the ground, with metal rods or conductors that transmit an electric current. These are separated by a fixed distance, but can be varied according to the needs of the designer. Some factors have also changed, such as resistance.

3.4. Soil Layers in the Pot

In nature, various factors contribute to soil drainage: gravity, percolation in all directions, evaporation due to sunlight on the surface, etc. Not to mention the abundant amount of soil, which in pots is limited to its size.

Due to the fact that the side walls of pots are impermeable and the hole in the bottom of the pot is the only way for water to drain out, important attention must be paid to preparing the soil to favour water drainage. This soil preparation favours water drainage and excludes a common problem with potted plants - root rot due to excess moisture.

To achieve this, the first step is to place polyester mesh at the bottom of the pot to prevent the pebbles from falling out, depending on the size of the pot's drainage hole. Then add a material that doesn't retain much water, such as expanded clay, coarse gravel, gravel, etc. Then a layer of sand on top of the expanded clay. Finally, place the top layer of fertilised soil until you think it's enough, as shown in figure 02.

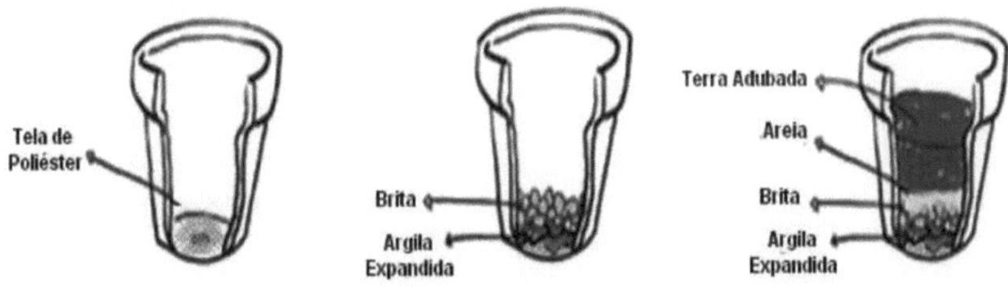

Figure 02 - Structuring the soil to facilitate water drainage in pots.

With this preparation, the fertilised soil, which has the highest water retention capacity, is in contact with factors such as ventilation and sunlight. Unlike the other lower layers, even though they are isolated by the impermeable walls of the pot and the factors mentioned above, due to the low water retention, it favours drainage into the drainage hole of the pot, thus preventing the lower regions of the soil from becoming waterlogged, and due to the preparation, they dry out evenly.

3.5. Sensor

Sensors are devices that change their behaviour under the action of a physical quantity and can directly or indirectly provide a signal that indicates this quantity. When they operate directly, converting a form of energy, they are called transducers. Those that operate indirectly change their properties, such as resistance, capacitance or inductance, under the action of a quantity, in a proportional way. The signal from a sensor can be used to detect and correct deviations in control systems and measuring instruments, which are often associated with open-loop (non-automatic) control systems, guiding the user (PELLISON, 2001).

There are characteristics of sensors that must be taken into account in a given situation.

Linearity is the degree of proportionality between the signal generated and the physical quantity. The more linear the sensor, the more faithfully it responds to the stimulus applied to it; these are ideal for systems that require precision. Sensors that don't have the linearity considered ideal are not necessarily discarded, but can be used in systems that don't require great precision and can work within a range. Sensitivity and response time are also important characteristics of a sensor (PELLISON, 2001).

3.5.1 Humidity sensors

There is currently a wide variety of humidity sensors, including: capacitive, resistive, thermal conductivity, optical and MEMS (FRUETT, 2008). In order to obtain quality when using a particular sensor, there are some characteristics that FRUETT (2008) considers desirable:

• High sensitivity;	• Low thermal drift;
• Quick reply;	• Long-term stability;
- Low cost;	- Durability;
- Small size;	- Reproducibility;
- Minimum hysteresis;	- Resistance against contaminants

3.5.2. Capacitive humidity sensors

Capacitive sensors are widely used to detect moisture in soils. They basically consist of two plates separated by a dielectric material, usually porous silicon, where the dielectric constant of the insulating solid between the plates can be changed by the moisture absorbed. The dielectric constant of water is high compared to most solids. Adding small amounts of moisture can produce large changes in the dielectric constant. In this way, the amount of moisture in materials can be detected by inserting electrodes into the material to be analysed. Capacitive sensors have more reliable results compared to resistive sensors, because the dielectric that separates the electrodes is part of the sensor body, so regardless of the soil the sensor is immersed in, the material separating the electrodes will always be the same, with only the humidity varying due to percolation between the dielectric and the soil (MOLIN, 2000).Percolation is the passage of water through a body by transmission from grain to grain, this phenomenon can be seen in figure 03.

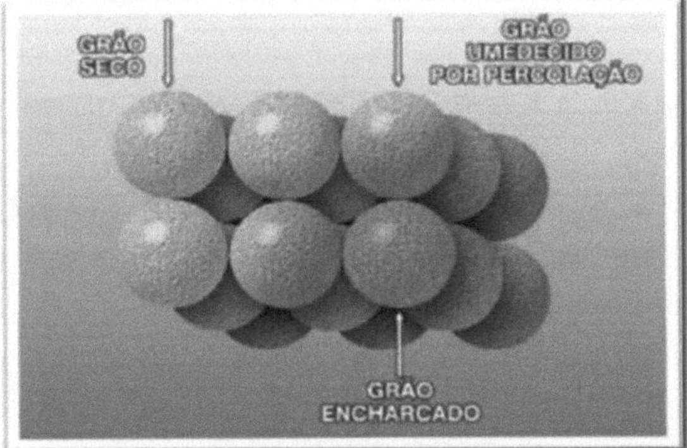

Figure 03- Demonstration of the transmission of water through a material by percolation.

Source: http://mundodaimpermeabilizacao.blogspot.com/2010/05/curso-basico-de-impermeabilizacao.html

MOLIN (2000) highlights the main characteristics of capacitive humidity sensors:

- Reduced thermal drift;
- Small size;
- Low resistance to contaminants;
- High response time.

3.5.3 Resistive humidity sensors

We currently have a wide variety of resistive sensors. The resistivity of materials can be used to detect the presence or variation of a wide variety of quantities. Whether it's a magnetic field, pressure or acceleration, chemical agents such as humidity, carbon monoxide, etc. (WILLIAMS, 1987).

There is also a wide range of resistive sensors known as chemical resistors. In all these components, resistivity is a function of the concentration of chemical agents present in the environment in which they are immersed. Chemical resistors are widely used to measure relative humidity, but can also be used for other purposes (WILLIAMS, 1987).

WILLIAMS (1987) mentions the main characteristics of resistive humidity sensors:

-Thermal drift   ;

-Small size;

-Low   resistance to contaminants;

-Hysteresis   ;

-Low   cost.

3.5.4 Reed Switch Magnetic Sensor

The Reed Switch is a magnetic field sensitive sensor. In its initial state, the two-pole switch is normally open (NO). When this device is passed through with an electric current, and is under the effect of a certain magnetic field, it closes its contact, allowing the current to pass through, functioning as a switch (FRANCISCO, 2004).

When installing the system, the direction of the magnetic field must be orientated correctly, perpendicular to the current or the contact. If the orientation is wrong, the switch may not close as expected, preventing the sensor from working correctly (FRANCISCO, 2004).

In figure 04 we can see the simple operation of the sensor by closing the circuit with the magnetic influence.

Figure 04 - *Reed switch,* without and with magnetic influence.

Source: http://www.designworldonline.com/articles/5779/315/Choose-Your-Best-Pneumatic-Cylinder-Sensor- Here.aspx

3.6   Microcontroller

Generally speaking, a microcontroller is an integrated circuit that contains, in a single package, a CPU, RAM memory, ROM memory and various peripherals such as: analogue-to-digital converters, timers, counters, communication ports and more.

Image03- PIC16F628A microcontroller.

Microcontrollers have components such as transistors and resistors in their internal structure, with a single crystal chip measuring a few millimetres.

squares. Each year more elements are added to them, making them more and more complete. Because of this, microcontrollers make projects cheaper and circuits more compact than those based on microprocessors, which usually require the use of several external components (ORDONEZ; PENTEADO; SILVA, 2005).

Thus, with the use of microcontrollers in process control, the desired data and variables can be processed by the control programme, installed in the microcontroller, according to the logical decisions pre-defined in this software. Mathematical comparisons can also be made that define the control functions (ORDONEZ; PENTEADO; SILVA, 2005).

According to (PEREIRA, 2005), the application of microcontrollers in projects brings several benefits, among them:

- Reduction in the number of components;
- Easy to implement complex control algorithms;
    - Reduction in the time needed to create new projects;
- Easy to change algorithms.

A microcontroller is usually small and cheap. The components are chosen to minimise size and be as economical as possible. So when you want a system to be easy, simple and cheap, a microcontroller is an excellent alternative.

3.6.1. PIC16F628A microcontroller

There is a wide variety of microcontrollers with different characteristics. So when designing a project, you should analyse the functions that a particular microcontroller has to offer, because it is not convenient to use a microcontroller with several integrated functions, when for a particular project you will only need to use one of them, thus generating unpleasant factors, for example:

- Higher consumption;
- Larger circuits due to the use of microcontrollers with unused ports;
- An unnecessary expense, since cheaper micros could be used.

The PIC 16F628A was used in this project because it is a simple, safe microcontroller with good immunity to electromagnetic interference and noise, FLASH programme memory, a high-performance RISC CPU and enough pins to be used for data input and output (PEREIRA, 2005).

Main features of the PIC 16F628A:

p It has 18 pins, 16 of which can be configured as data input and output;

-2k of memory for 14-bit instructions;

-128   bytes of EEPROM memory;

-224   bytes of RAM;

-Each I/O pin capable of supplying or consuming 25mA;

-A   set of 35 instructions;

-Possesses   serial USART, necessary for transferring data to the radio transmitter;

*State change interrupt   on pins RB4 to RB7 (INT_RB).

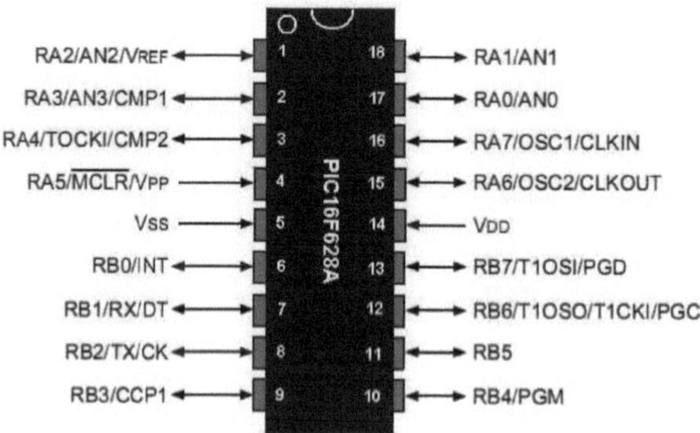

Figure05 - PIC 16F628A microcontroller and its respective pins.

3.7.   Operational Amplifier

The Operational Amplifier (AOP) is a component constructed from the junction of resistors, capacitors and transistors in the same body. This component was once very used to compute mathematical operations such as sums and integrations. This is why it is called an Operational Amplifier. As technology advanced, the name Operational was added to its name due to its versatility in previously complex implementations and in the most varied projects (ALBUQUERQUE, 2009).

The graphical representation of the 741CN operational amplifier is shown below in figure 06:

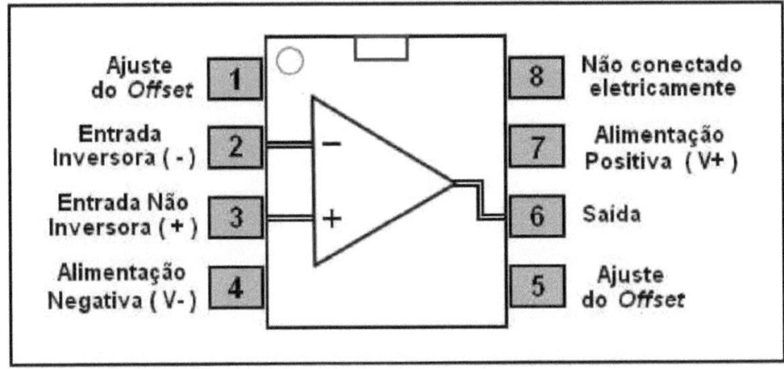

Figure 06 - Pinout of an Operational Amplifier.

An AOP is an amplifier with very high gain. It has two input terminals: one called the negative inverting terminal and the other called the positive non-inverting terminal. The output voltage is the difference between the two inputs, multiplied by the open-loop gain (MALVINO, 1997).

It is very difficult to list all the applications of this fantastic component, but we can say that it is used in most industrial control system equipment, nuclear and petrochemical instrumentation, medical equipment, computers, etc. (MALVINO, 1997).

The main characteristics of an AOP are:

-Infinite input resistance;

-Nil   output resistance   ;

   -Infinite voltage gain   ;

   -Infinite frequency response   ;

   -Temperature insensitivity   .

There are polarisation methods for an operational amplifier:

s Without Feedback;

• Positive feedback;

• Negative feedback.

3.7.1. AOP Without Feedback

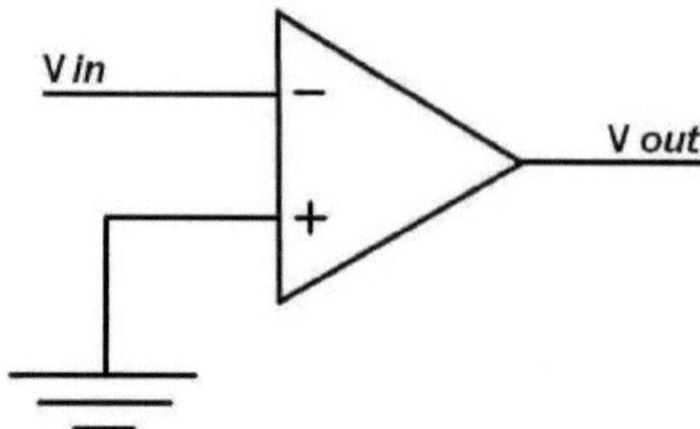

Figure 07 - AOP without feedback.

This mode is known as open-loop operation because it uses the gain of the operational unit stipulated by the manufacturer, i.e. there is no control over it. This mode of operation is widely used in comparator circuits (MALVINO, 1997).

3.7.2. AOP as a comparator circuit

Among the many basic circuits that use AOPs, there is the comparator circuit, which is the working principle of the sensor in this project. A comparator is a circuit with two input voltages (non-inverting and inverting) and one output voltage. When the non-inverting voltage is higher than the inverting voltage, the comparator produces a high voltage; when the non-inverting input is lower than the inverting input, the output goes low. The high output symbolises logic level "1" and logic level "0" will be the lowest (ALBUQUERQUE, 2009).

CHAPTER 4

WORK METHODOLOGY

The aim of this project is to manage and maintain the optimum level of soil moisture, using a resistive sensor (electrode) integrated into the microcontrolled system and a motor-pump as an actuator. Developed in a miniaturised form specifically to meet the water needs of plants grown in restricted containers. It can also be used in small gardens and conservatories. Its demand can be easily increased according to the needs of the crop by using pump motors with greater flow and power, as it is isolated from the circuit by a relay.

The system activates a relay by switching on the pump motor, automatically starting irrigation if the soil moisture level is below the minimum level stipulated and there is water in the reservoir. Likewise, it deactivates the relay if the soil moisture level exceeds the stipulated maximum value or if there is no water in the reservoir. All process statuses are illuminated by red and green LEDs.

	ESTADO DO RESERVATÓRIO	ESTADO DO SOLO	ESTADO DO MOTOR
1º	Cheio	Umido	Desligado
2º	Cheio	Seco	Ligado
3º	Vazio	Umido	Desligado
4º	Vazio	Seco	Desligado

Figure 10 - System states indicated by LEDs.

As shown in figure 10, there are four situations and only in one will the pump motor be switched on, so that:

i 1st situation - Green tank LED indicates that the tank is full. Green ground LED informs you that the ground is damp (Note: Motor switched off);

- 2nd situation - Green LED on the tank tells you that it is full. Red LED together with the flashing green LED on the ground informs you that the ground is dry and is being irrigated (Note: Motor switched on);
- 3rd situation - Red LED on the tank indicates that it is empty. Green ground LED informs you that it is wet (Note: Motor switched off);
- Situation 4 - Red LED on the tank tells you that it is empty. Red ground LED informs you that the tank is dry (Note: Motor does not start as there is no water in the tank).

1.1. Circuit Description

1.1.1. Power supply

The system has two types of power supply:

- A 220 volt alternating power supply for the pump motor;
- A rectified and filtered 05 volt supply for the circuit and sensors.

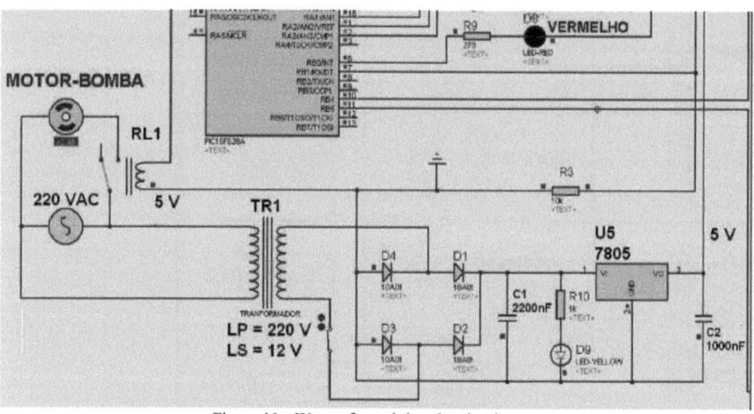

Figure 11 - Ways of supplying the circuit.

4.1.2. Soil sensor

Focussing on plants grown in restricted environments with small containers, a resistive sensor was developed where its contact area with the ground was also reduced, making it possible to use it even in situations with very small containers.

The following materials are used to make the electrode: a copper-plated phenolite plate with a rectangular shape approximately four and a half centimetres high by two centimetres wide. This plate was corroded with iron perchloride to create two isolated tracks. Wires large enough to carry the information from the electrode to the sensor circuit were then soldered to each track. Silicone was used where the welds were made, isolating the welded area from the humidity in which the electrode is immersed, thus preventing oxidisation. The electrode in contact with the sensor's ground was designed to be small, with a fixed contact distance on the same body, as shown in image 04 below.

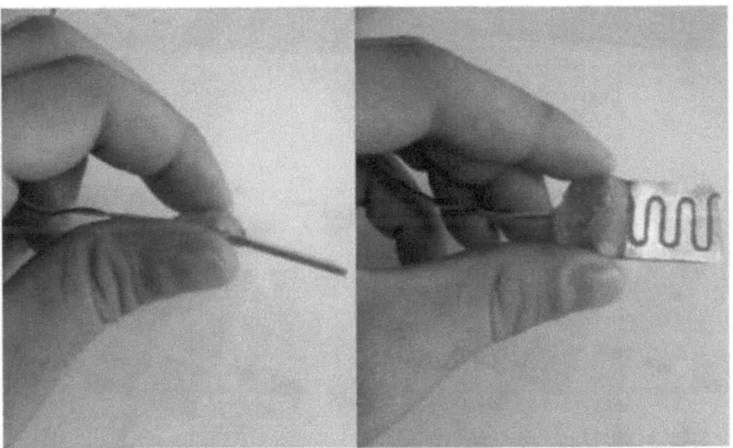

Image 04 - Electrode, part of the sensor that is in contact with the ground.

The sensor electrode is installed in the soil at the effective depth of the roots. When the soil is dry, its resistance increases, making it difficult for current to pass through, indicating when to irrigate. As the water is absorbed, the resistance of the soil decreases, allowing current to pass between the electrodes and thus closing the circuit.

The operation of the sensor is based on the use of two AOPs as comparators, strategically interconnected so that as the current detected by the electrode varies, the AOP compares the inputs and varies their outputs respectively.

In order to be able to work at a humidity level varying between a given humidity range, it was necessary to use two AOPs. This means that the circuit will only activate the motor-pump when the current level in the electrode, which is proportional to the humidity, is 20%, and it will remain switched on until it reaches 80%. Likewise, it remains switched off until it returns to the 20% level. This prevents the sensor from switching on and off in a disorderly fashion.

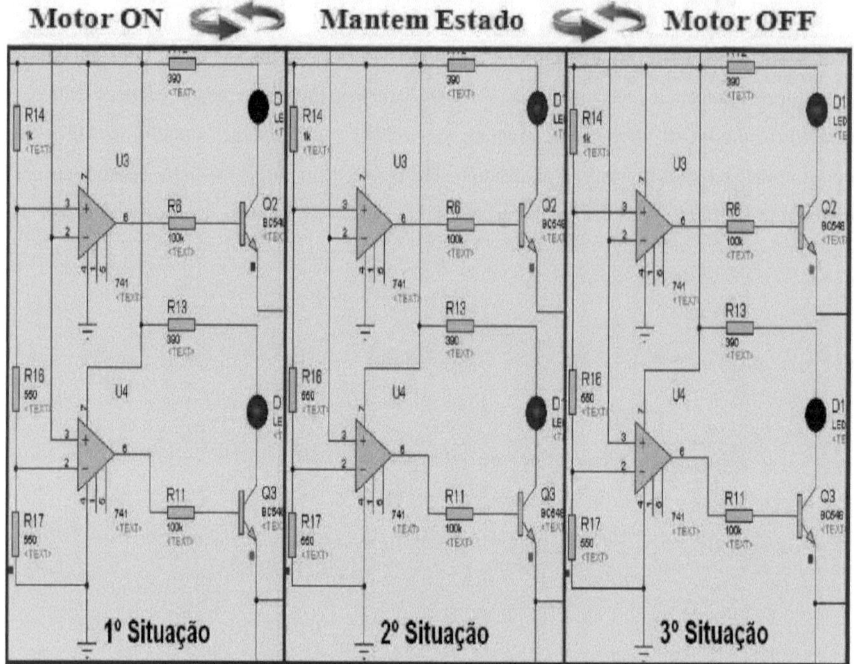

Figure 12 - Possible sensor states.

As shown in figure 12, when the sensor is in the 1st situation, the system understands that the soil is moist at the maximum level, disabling the pump motor. As the soil loses moisture, it moves to the 2nd situation, remaining switched off until it reaches the 3rd situation, where it understands that the soil is at the minimum moisture level, enabling the pump motor. At this point, irrigation begins, so that the soil gains moisture and returns to the 2nd situation, remaining switched on until it reaches the 1st situation, switching off the pump motor and starting the whole process.

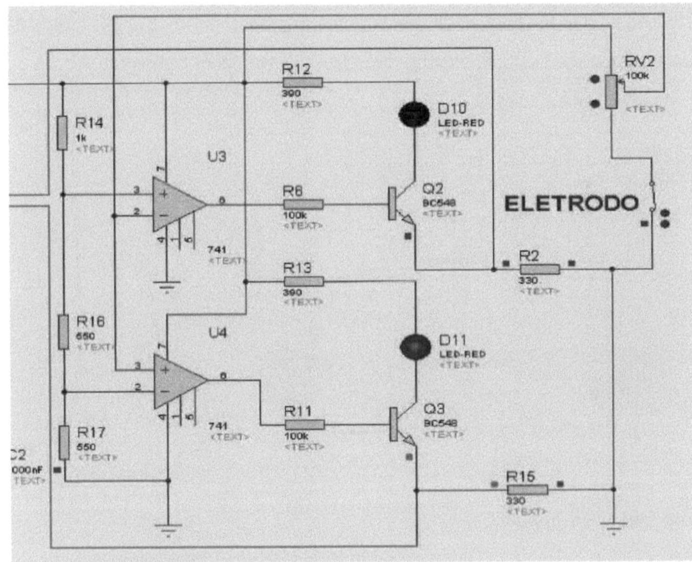

Figure 13 - Sensor circuit indicating dry soil status.

A 100 k potentiometer is placed in series with the electrode to allow the system user to vary the resistance and therefore the sensitivity of the sensor. This is so that the variations caused by different types of soil can be minimised. This also provides better conditions for plants that naturally require more water.

Image 05 - Potentiometer in series with the electrode and next to it its picture in the circuit.

4.1.3. Water Tank Sensor

The system recognises the absence of water in the tank due to a magnetic reed switch sensor at the bottom. On one axis, there is a magnet with a floating material (Styrofoam), and when

25

the water is used, the Styrofoam next to the magnet descends following the water level and is directed along an axis until it approaches the reed switch, closing the circuit, as shown in figure 14.

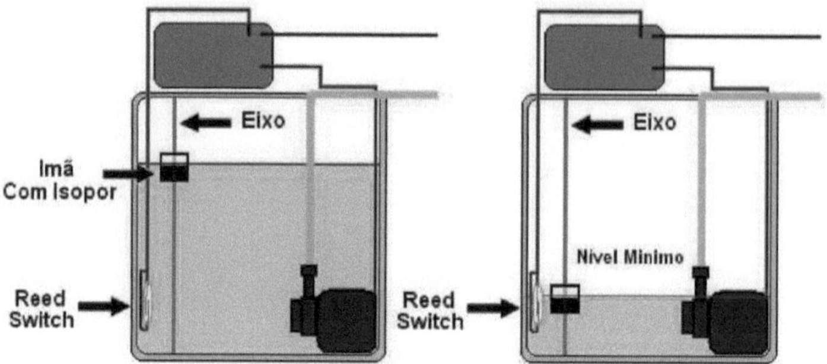

Figure 14 - Water recognition system in the reservoir.

4.1.4 Microcontroller Pins and Their Functions

When programming, we can define functions for sixteen of the microcontroller's pins, which can be configured as data input or output. In this project, only eight pins were used. The following functions were assigned.

As input data:

-RB1   ,pin 7 -Servo sensor;

-RB4   , pin 10 - Ground sensor;

-RB5   , pin 11 - Ground sensor.

As output data:

-RB0   , pin 6 - Lights up red reservoir LED;

-RA3   , pin 2 - Lights green reservoir LED;

-RA2   , pin 1 - Lights red ground LED;

-RA1   , pin 18 - Green ground LED lights up;

-RA0   , pin 17 - Saturates the transistor that switches on the relay.

Below is the flowchart of the variation in the logic level of the pins duly programmed and configured as outputs of the microcontroller, by means of variations in the inputs.

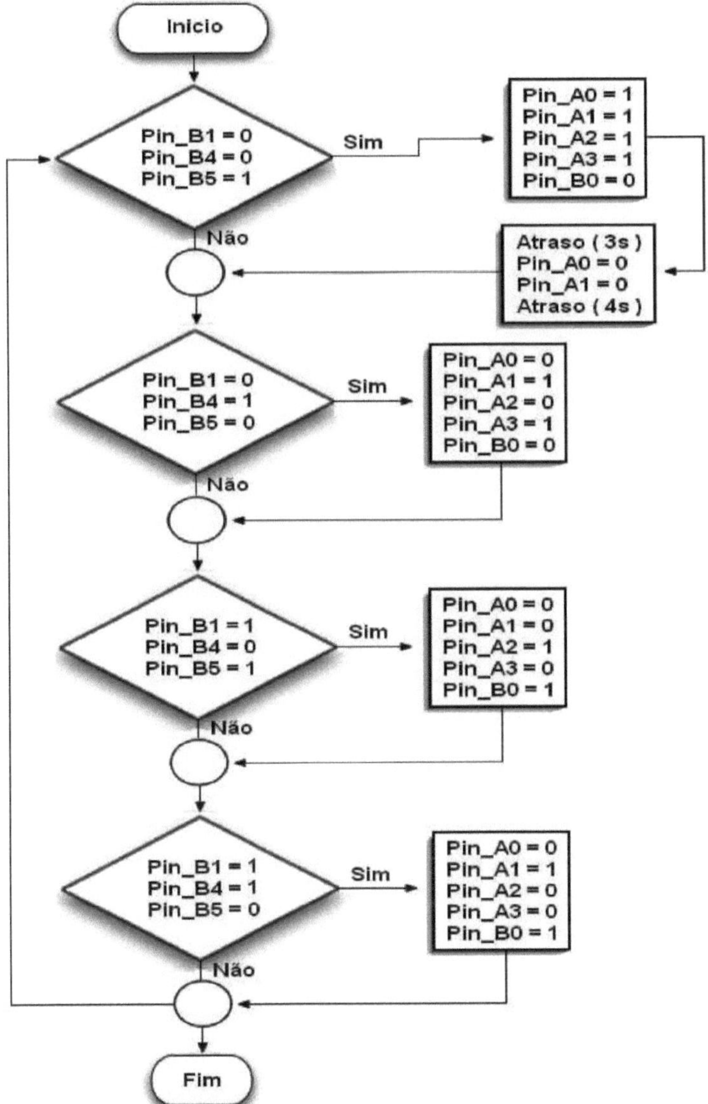

Figure 15 - Flowchart of the logic level of the output pins, using the inputs.

4.1.5 Pump-motor drive

The drive starts **when pin 17 of the RAO changes from logic level "0" to logic level "1",** **saturating the transistor** feeding the relay coil and closing the circuit that powers the motor-pump.

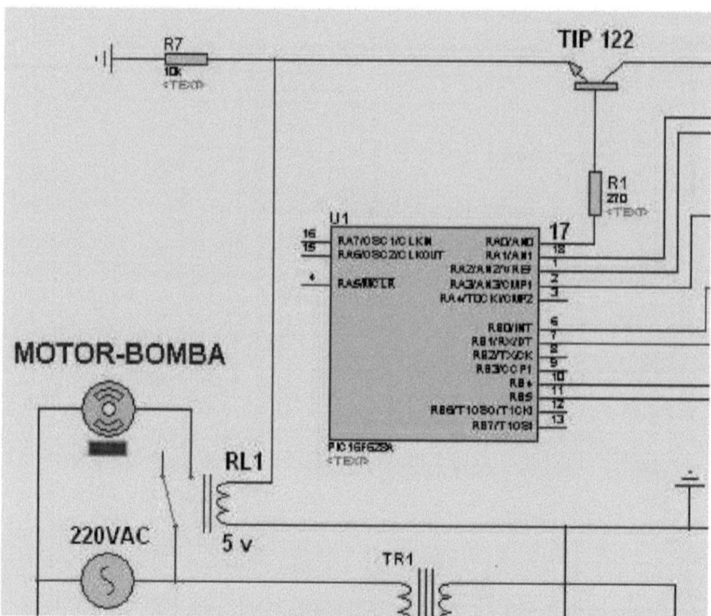

Figure 16 - Pump-motor drive connections.

1.2.   Software

The system management software was programmed in the C language. It was used because it is a language that allows for great speed in creating projects and because it is easy to adapt to changes from one system to another.

A programme capable of interpreting the sensor signals via the microcontroller's parallel ports was developed, sending commands to the actuator (Motor-Pump) when necessary.

The software understands the variations in the signals obtained from the soil and reservoir sensors, which are connected to pins configured as inputs. When they are at the levels defined as the time to irrigate, the system automatically enters a routine to start the irrigation process. This routine makes the pump motor run in a three-second off, four-second on cycle, avoiding wasting water by giving it more time to infiltrate and recognise the sensor, avoiding wear and tear due to leaching, which weakens the soil due to the excessive passage of water carrying nutrients with it.

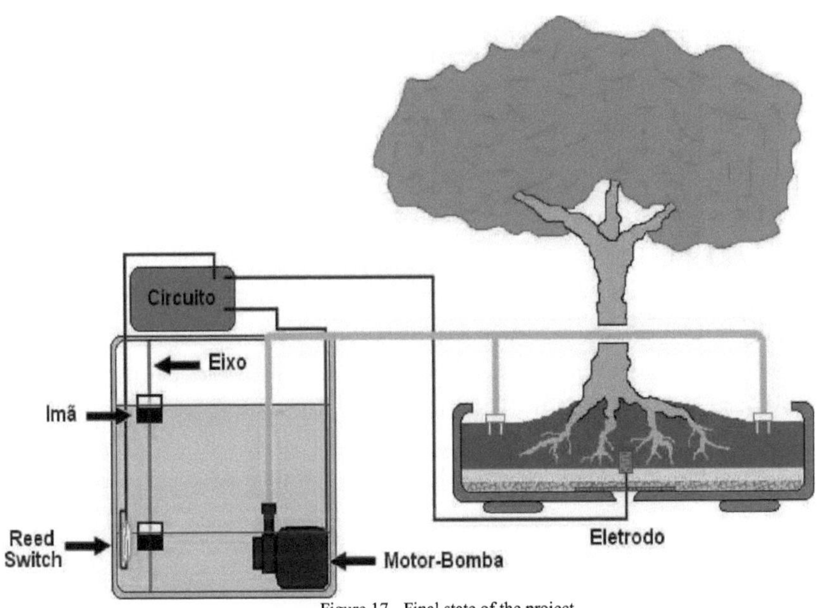

Figure 17 - Final state of the project.

1.3. Difficulties in Project Development

Calibrating the sensor was the biggest difficulty encountered in developing this project. To regulate its sensitivity, it is first necessary to define the contact area and the distance between the electrodes. To avoid variations in conductivity related to changes in the distance between the electrodes, an electrode with two contact poles on the same body with a fixed distance isolated from each other was developed. Once this was defined, based on practical tests, the resistances in the sensor were varied until the desired result was achieved.

CHAPTER 5

RESULTS

Using a small reservoir in which the motor-pump was submerged, the electrode in contact with the soil was in place and the water-directing pipes were fixed at specific points in the soil of the pot. Possible situations were realised in an induced manner, whereby when the magnet was removed from the reed switch magnetic sensor, the system started irrigation automatically, and in a few seconds the electrode identified the humidity level specified by the potentiometer, thus also automatically stopping the motor-pump, reacting identically to simulations carried out virtually on a circuit simulator, confirming its viability in practice.

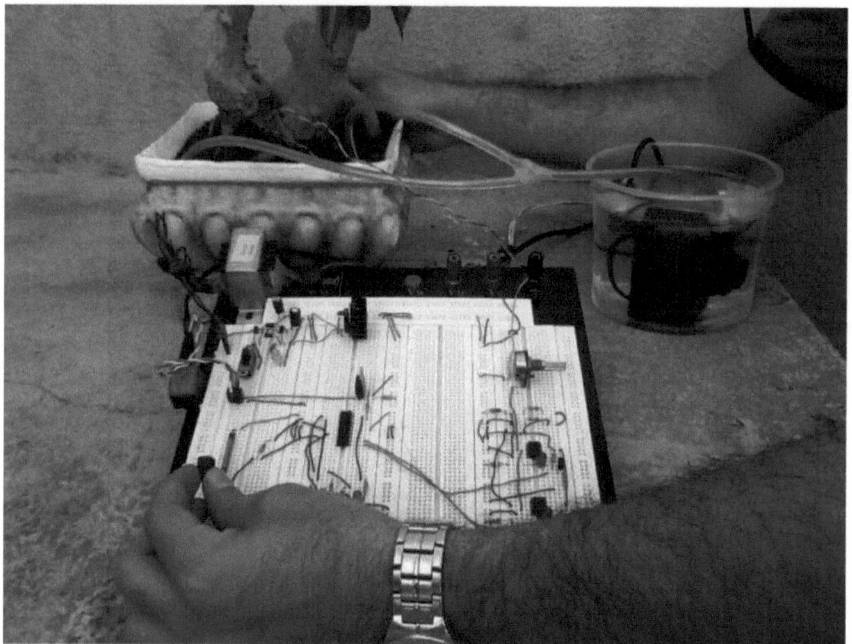

Image 06 - Removing the magnet from the vicinity of the reed *switch* magnetic sensor, so that the system understands that there is water in the tank.

CHAPTER 6

CONCLUSION

Electronics play a fundamental role in modernising various activities, making it possible to automate equipment, making activities faster and more precise.

This work shows the physical and chemical characteristics of soil and their influences. It provides a standard way of preparing soil in pots, in order to achieve uniform drainage and, above all, the ideal amount of water for healthy plants.

Finally, based on the practices carried out, we realise that the objective for which the project was developed has been achieved. With the creation of a portable, automated irrigation system that is simple and easy to adapt to the different situations normally found in the type of crop it was developed for. Adding a range of theoretical and practical knowledge, it can be easily implemented with other technologies that can extend its functionalities. As a result of this project, a higher level of safety, comfort and flexibility has been achieved. With the use of this equipment, people who grow crops at home can be away for longer periods of time, eliminating the need for the user to have a responsible person to water the plants correctly during periods of absence due to long journeys, holidays and so on.

5.1. Future implementations

To implement new functions in future work, a suggestion could be to add a message sending and receiving system to the system, making it possible to use this technology to remotely monitor, receive information and even allow the user to interfere in the process regardless of where they are.

CHAPTER 7

REFERENCES

ALBUQUERQUE, R.O.; SEABRA, A.C. **Using electronics with AO, SCR, TRIAC, UJT, PUT, CI 555, LDR, LED, FET and IGBT**. São Paulo: Erica, 2009.

OPERATIONAL AMPLIFIER. Available at : <http://www2.eletronica.org/apostilas-e-ebooks/componentes/AOP.pdf>. Accessed on 08 June 2011.

BRADY, N.C. **Nature and properties of soils**. 7th ed. Rio de Janeiro: Freitas Bastos, 1989.

BRAGA, N. C. **Basic electronics course**. São Paulo: Saber, 1999. v. 1, 140 p.

FRANCISCO L.; Marcia C.S.M.L.; FREDERICO A.O.C. **Using magnetic sensors on an air rail. Revista Brasileira de Ensino de Física**, Universidade Estadual do Rio de Janeiro, v. 26, n. 3, 233 - 236 f.. 2004.

FRUETT, F. **Integrated Silicon Sensors IE012, Humidity Sensors**. UNICAMP - FEEC - DSIF. São Paulo, Campinas, 2008.

IUBlog. **Gardening - Water the Plants**, 20 January 2011. Available at: <http://blog.institutouniversal.com.br/index.php/2011/01/regue-as-plantas/>. Accessed on 09 June 2011.

LEPSCH, Igor F. **Formação e Conservação dos Solos**. São Paulo: Oficina de Textos, 2002.

LOPES, Alfredo Scheid (transl. and adapt.). **Manual de fertilidade do solo**. São Paulo: ANDA/POTAFOS, 1989.

MACHADO, P. L. O. A. et al. **Mapping of electrical conductivity and relationship with clay of Lat os soils under no-tillage.** Pesquisa Agropecuária Brasileira, v.41, n.6, Brasília, June 2006.

MALVINO, Albert Paul. **Electronics**. Volume 2. 4ª . Edition. São Paulo: Pearson Makron Book, 1997.

McBRIDE, R. A; GORDON, A. M; SHRIVE, S. C. **Estimating forest soil quality from terrain measurements of apparent electrical conductivity**. Soil Science Society of American Journal,v.54, p.255-260, 1990.

MOLIN J.P.; GIMENEZ L.M.; PAULETTI V.; SCHMIDHALTER U.; HAMMER J. **Measurement of Soil Electrical Conductivity by Induction and Its Correlation with Production Factors**. Eng. Agrícola, Jaboticabal, v.25, n.2, p.420-426, May/Aug. 2005.

MOLIN, J. P. **Precision Agriculture Journal**. Esalq/USP, Brasília (DF), March 2006.

MONIZ, A. C. (Coord.). **Elements of pedology**. São Paulo: Technical and Scientific Books, 1975.

ORDONEZ, E. D. M.; PENTEADO, C. G.; SILVA, A. C. R. **MicrocontrollersFPGAs: Applications in Automation**. São Paulo: Novatec, 2005.

PELLISON, Antônio T. **Proposal for an automated system to control the water table and sub-irrigation**. Master's Degree in Irrigation and Drainage, Faculty of Agronomic Sciences, UNESP. São Paulo, Botucatu, 2001.

PEREIRA, F. **Microcontrolador PIC - Programação em C**. 3. Ed. São Paulo: Érica, 2005.

QUEIROZ, T. M. **Development of an automatic system for precision centre pivot irrigation**. 2007. 141 f. Thesis (Doctorate in Agronomy) - Luiz de Queiroz College of Agriculture, University of São Paulo, Piracicaba, 2007.

QUEIROZ, T. M. et al. **Development of software and hardware for precision irrigation using centre pivot**. Engenharia Agrícola, Jaboticabal, v. 28, n. 1, p. 44-54.

REICHARDT, K. **Water in agricultural systems**. São Paulo: Manole, 1987. 188p.

SILVEIRA, **Paulo Rogério da Automation and discrete control**. São Paulo: Érica, 1998. - Study and Use Collection. Industrial Automation Series. 228p.

SOARES, Vinícius. L. A. **Automation of an irrigation system using the bubbler method**. 2010. 48 f.. Monograph (Degree in Industrial Mechatronics) - Federal Institute of Education, Science and

Technology of Ceará, Cedro campus, 2010.

YOSHIOKA M.H.; LIMA M.R. **Solo na Escola University Extension Project** - Federal University of Paraná - UFPR Department of Soils and Agricultural Engineering. Curitiba, 2005.

WILLIAMS, B. G.; HOEY, D. **The use of electromagnetic induction to detect the spatial variability of the salt and clay content of soils**. Australian Journal of Soil Research, v.25, n.1, p.21-7, Melbourne, 1987.

JAYNES, D. B. **Improved soil mapping using electromagnetic induction surveys**. In: P. C. ROBERT, R. H. RUST and W. E. LARSON. International Conference on Precision Agriculture. Minneapolis, p.169-79, 1996.

ANNEX I

Detailed microcontroller programme in C language.

```c
#include <16f628a.h>
#use delay(clock=4000000)
#fuses INTRC_IO,NOWDT,PUT,NOBROWNOUT,NOMCLR,NOLVP

void main()
{
output_a(0);
output_b(0);
while(true)
{
if (!input(pin_b1) && !input(pin_b4) && input(pin_b5))// Reservatório cheio, terra seca. Motor ligado.
{
output_high (pin_a3);// O led verde do reservatório ligado.
output_high (pin_a2);   // O led  vermelho do solo ligado.
output_low (pin_b0);// O led vermelho do reservatório desligado.
output_high (pin_a0);// Pino que aciona o relé que liga o motor ligado.
output_high (pin_a1);// O led verde do solo ligado.
delay_ms(3000);//  3 segundos.
output_low (pin_a0);// Pino que aciona o relé que liga o motor desligado.
output_low (pin_a1);//  O led verde do solo desligado.
delay_ms(4000);//4 segundos.
}
if (!input(pin_b1) && input(pin_b4) && !input(pin_b5)) // Reservatório cheio, terra umida. Motor desligado.
{
output_low (pin_a0);// Pino que aciona o relé que liga o motor desligado.
output_high (pin_a1);// O led verde do solo ligado.
output_low (pin_b0);// O led vermelho do reservatório desligado.
output_high (pin_a3);// O led verde do reservatório ligado.
output_low (pin_a2);// O led  vermelho do solo desligado.
}
if (input(pin_b1) && !input(pin_b4)&& input(pin_b5)) // Reservatório seco, terra seca. Motor desligado.
{
output_low (pin_a0);// Pino que aciona o relé que liga o motor desligado.
output_low (pin_a1);// O led verde do solo desligado.
output_low (pin_a3);// O led verde do reservatório desligado.
output_high (pin_a2);// O led  vermelho do solo ligado.
output_high (pin_b0);// O led vermelho do reservatório ligado.
}
if (input(pin_b1) && input(pin_b4) && !input(pin_b5)) // Reservatório seco, terra umida. Motor desligado.
{
output_low (pin_a0);// Pino que aciona o relé que liga o motor desligado.
output_low (pin_a2);  // O led  vermelho do solo desligado.
output_high (pin_b0);// O led vermelho do reservatório ligado.
output_low (pin_a3);//  O led verde do reservatório desligado.
output_high (pin_a1);// O led verde do solo ligado.
}
}
}
```

ANNEX II

System peripherals: power supply, relay, pump motor and ground electrode connected to the circuit.

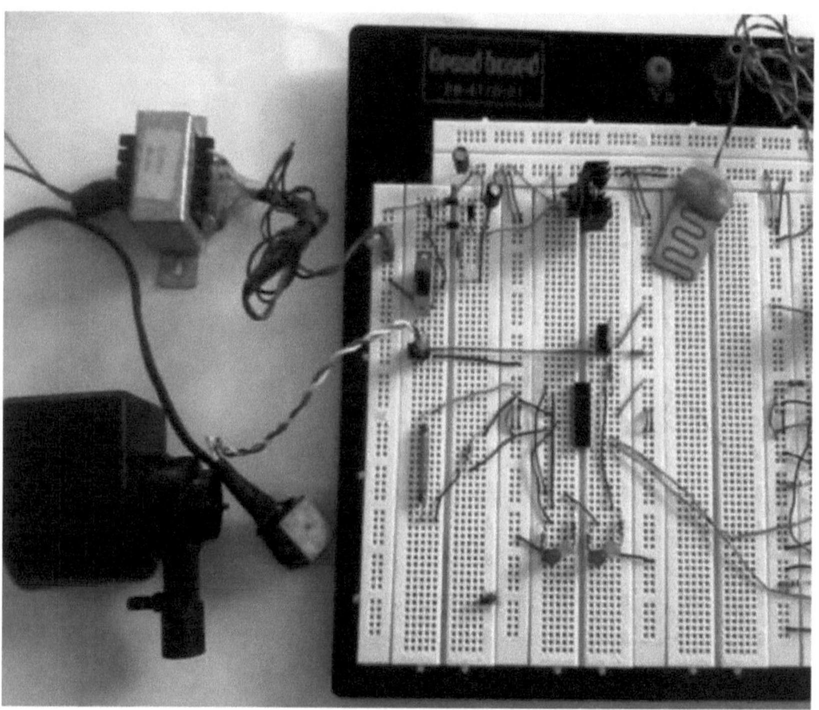

ANNEX III

Closer visualisation of the complete circuit.

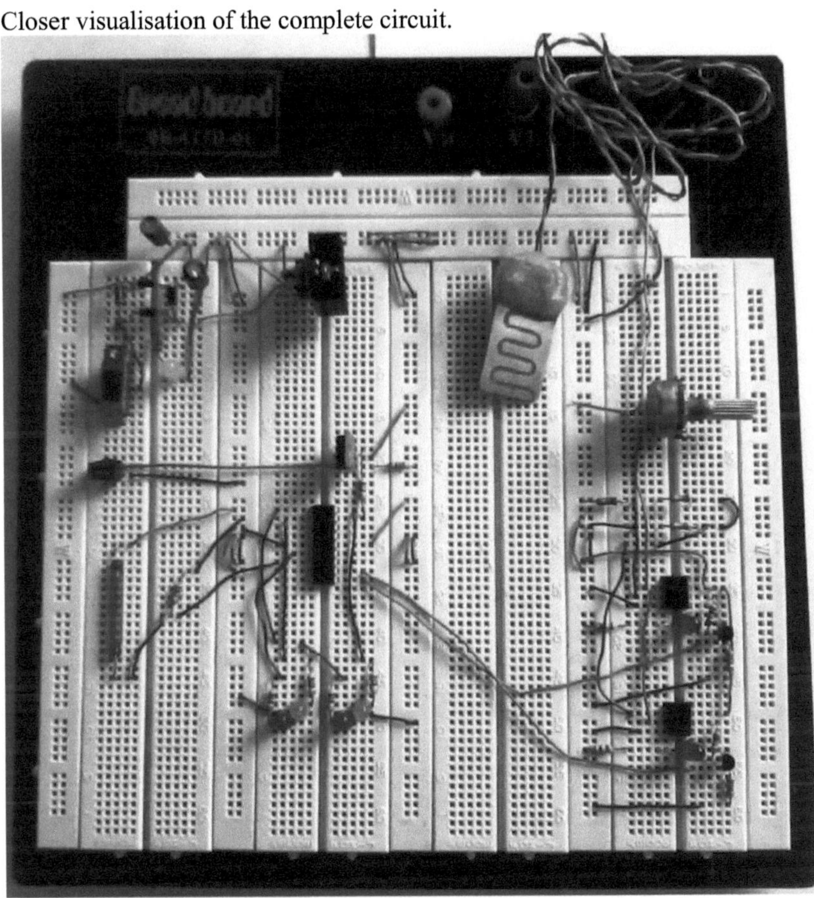

ANNEX IV
Circuit schematic diagram.

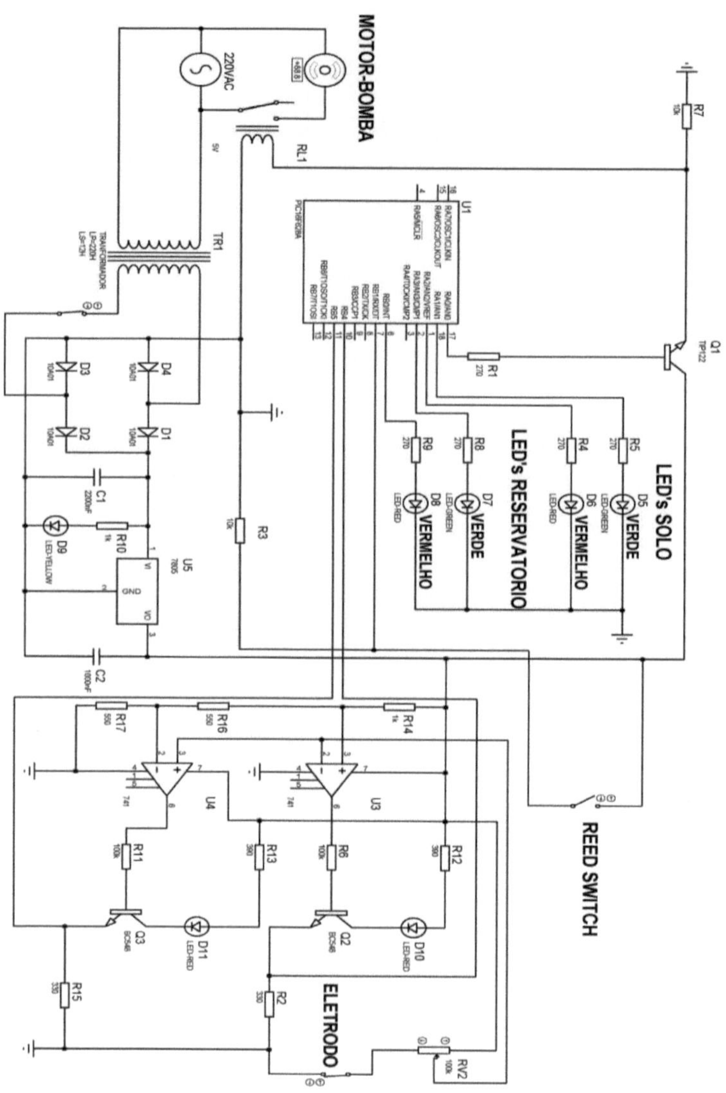

ANNEXV

Illustration of the system properly installed for automatic soil moisture monitoring.

The image below shows the adjustments made to the sensor's sensitivity, varying the potentiometer in order to achieve a lower percentage of humidity, since the specimen used in this case is a Bonsai of the Fícus species, a plant that is more prone to root rot if it is grown in environments where there is constant excess humidity.